ANIMAL RESCUE

A guide to rearing some
of the more common species
of Australian marsupials

by
Ken White

Cover and illustrations by Wolfgang Grasse and Michael Searle.

ANIMAL RESCUE

A Guide to rearing some
of the more common species
of Australian Marsupials

This book is dedicated to my wife, Joy, without whom it could not have been written.

"The beauty of God's world (is) given for the enjoyment of all creatures." - Tolstoy (Resurrection).

"Animals are such agreeable friends; they ask no questions; pass no criticism." - George Eliot.

ISBN 0 949457 73 6

Published by
REGAL PUBLICATIONS
24 Wellington Street, Launceton, Tasmania
Telephone (003) 31 4222

CONTENTS

INTRODUCTION

This book is the result of my wife's seven years experience in rearing and looking after the more common species of orphaned and injured wildlife. During that time I was persuaded on many occasions to share our bed with baby possums and wallabies, once, even, with a Tasmanian Devil. But it was a labour of love for Joy who went through many periods of frustration as she battled to save a sick joey or an injured animal. Initially, of course, joeys are totally reliant on their newly-found parent and can be demanding as a new-born child. Sometimes, one will grow up and, in human terms, become an 'unforgettable character.' We had a couple that I'll certainly never forget. For marsupials, with which this book is concerned, exhibit many of the traits of homo sapiens - affection, possessiveness, jealousy, cheekiness and so on. However, getting the babies through the first couple of weeks isn't easy. It requires time, patience and dedication - and knowing the correct formulas and food to feed them. The formulas given throughout this book have proved very successful after, unfortunately, early trial by error. Most of the youngsters have not only survived but have grown into healthy adults to be released back into their natural habitat. I would like to take this opportunity to give special thanks to Dr. Brian Larner, of Devonport, for his advice and suggestions, and to the Tasmanian Fauna Society and its secretary, Ellen Jensen, and to Mr. S.E. (Hans) Wapstra, of the Tasmanian Department of Parks, Wildlife and Heritage.

Ken White,
Sisters Beach,
Tasmania,
1993

PREFACE

When I knew Joy and Ken White in Darwin in the late nineteen-seventies, their interests appeared to centre on good food, gardening and current affairs. Ken was a court reporter with the Northern Territory News and Joy later became a clerk at the Royal Darwin Hospital. They sometimes talked about returning to 'down south,' Sydney for them, but I didn't expect them to be in a hurry.

Yet, in 1985, they wrote from a farm in Tasmania and I wondered how long they .would endure the cold climate. A year later I heard of Joy's crusading efforts to rescue young marsupials from certain death and within a few years their farm had become a wildlife orphanage, with young animals being brought to them by animal lovers and animal hunters alike.

It is a wonderful, miraculous thing to see a passionate pursuit take shape because of the recognition of the ancient bond between humans and creatures from the wild. Even when Joy and Ken try not to become attached to an animal, because eventually it was to be released, they nevertheless care for it with total devotion as long as it needs to be a member of their household. This extends the bond to selfless love and true compassion.

In the course of seven years they have returned more than 100 animals to the wild. Some of them they know have reproduced. It may seem a drop in the bucket in the face of the figures of wildlife fatalities in Tasmania, but at some point in the future it could be that the prodigy of an animal reared by Joy and Ken will be the deciding factor in whether a species becomes extinct or survives in Tasmania.

This small book contains only a few of hundreds of stories Joy and Ken could tell about their life with wildlife. But it contains all the vital information gathered by Joy about rearing, feeding and healing of helpless orphans from the bush.

Although it would be good to have a sympathetic veterinarian and a National Parks and Wildlife Office nearby, any animal lover willing to take on the responsibility can rear young Australian marsupials with the help of this handbook and a few products obtainable at the chemist and greengrocer.

In this decade, public awareness is growing that time is running out if we want to stop and turn the decline of Australian wildlife populations. Therefore, this book should have wide appeal, among the general reading public, in schools, and especially in country areas.

This book can also help to make knowledge of the rearing of Australian marsupials as general and widespread in the nineties, as the knowledge of raising Australian native vegetation has become in the eighties. It is the logical next step.

It is my hope that by the year 2000 this handbook will sit on the bookshelf of most households in the country.

Lolo Houbein.
Author, and Co-founder of Trees for Life
(South Australia's branch of Men of the Trees).

CHAPTER 1
The Young Joey

There is an appalling mortality of hand-reared orphaned wildlife, especially among kangaroos and wallabies. The basic problem lies in the difficulty of replicating the pouch environment for the joey. Marsupials - and there are 249 living species, not all Australian - give birth to minute, hairless babies. Gestation periods range between 12 and 36 days which means pregnancy is short, even for the largest 'of them. Most are seasonal breeders and their reproductive cycles are timed so that the young finally leave the pouch at the most favourable time of the year. Newly-born joeys instinctively crawl from what is called the cloaca to the pouch where they attach themselves to a nipple. At least one third of their pouch life is attached to the mother in this way. They are, however, unable to regulate their own body temperature. This thermoregulatory capacity develops during pouch life and is only functioning effectively when the joey is about to leave its mother and make its own way in the world. Joeys still attached to the teat usually die quickly after separation. Those cared for as orphans generally are in the last half of pouch life.

Another major problem in rearing joeys is matching the milk. The composition of marsupials' milk varies through lactation, and it's quite a sophisticated system. Macropodids (kangaroos and wallabies) can be suckling a pouch-dependent joey, as well as an older joey at 'foot', on two completely different types of milk from different mammary glands The high lactose content in cow's milk, so often mistakenly given to young joeys, causes diarrhoea leading, in most cases, to death. The baby simply cannot digest the lactose. Similarly, the addition of sucrose to milk also causes diarrhoea. But once the joey begins eating a significant amount of grass, the effect of which stimulates development of a microbial fermentation in the stomach, 'the runs' cease to be a problem as the sugars are fermented before reaching the small intestine. This probably all seems pretty involved, but it's important in understanding the problems involved in rearing Australian wildlife. Most of the milk formulas given in the following pages include Di-Vetelac as the main ingredient. It is a commercial product, low in lactose and available from vets.

More success has been achieved with Di-Vetelac than with other preparations. There are however, substitute milks, made by a South Australian company and recommended by some vets, called Wombaroo. Reference is made to their products later.*

Because of the young joey's inability to keep warm if it's taken from the mother's pouch, it's essential to provide it with adequate covering and constant warmth. If it's hairless, the chances of rearing it are pretty slim. Should the animal be in this condition, immediately rub its body with a top grade baby oil to prevent loss of moisture, and keep it at a steady temperature of 36 deg C. Once it's covered with hair, a temperature of between 30 and 32 deg (maximum) should be maintained. if it has any scratches or cuts, rub a little Ungvita on them.

Don't use hot water bottles, as they cool too quickly. Better to place the orphan in a substitute pouch on a waterproof electric blanket, or near a room heater or combustion stove. If it is near a heater or stove, place a container with water close by as this will provide the necessary moisture to prevent dehydration. Special electric pet blankets, as well as heated slippers, are available from some vets, but if you use them check the joey regularly to see it's not overheating. This can easily occur. When using electric blankets, don't forget room temperature must be taken into account when regulating the thermostat. Pouches are easily run up. Old jumpers or pillow cases lined with something soft and warm will suffice. The joey will pop its head through the neck of the jumper as it would out of a pouch. But whatever material is used, make sure it's washable, and the pouch itself is fairly shallow. When the joey is inside, and you're not carrying it attach the pouch to the side of a chair, or box, as the animal will prefer this position to lying flat on its side. When carrying the joey around, place it in a pouch against the chest and under a jumper (if the weather is cool). This will enable the youngster to attune itself to human smells. As it gets older, it should be allowed some exercise and the opportunity to leave its pouch and explore a sheltered area outside the house. Encourage it to lap as soon as it can stay out of the pouch long enough, as this will help prevent aspiration pneumonia.

Native animals in captivity are extremely vulnerable to infection and have relatively undeveloped immune systems, particularly when young, so scrupulous cleanliness is essential. Sterilise all utensils used in the preparation of the milk formula by boiling them. Use marsupial teats and feeding bottles, available from vets, but continue to sterilise them until the animal is eating substantial amounts of solid food, especially native grasses with soil attached to the roots. Except in an emergency, don't use an eye-dropper for feeding, as it can damage sensitive mouths and set up a fatal infection. Cooled boiled water, too, should be used in the preparation or the formulas.

After reading this, you'll have some idea of what you'll be letting yourself in for if you become a foster parent to a young joey. In the next chapter I'11 explain how to go about keeping them alive. But a word of warning: you cannot rear wildlife if you have a cat. Or, if you do own a cat, they must be kept well apart. The reason is that cats carry a disease which is fatal to most marsupials. It's called toxoplasmosis and it's dealt with more Fully in Chapter 13.

* NEW MILK FORMULA

Since the writing of this book, a new milk formula which the author and his wife found to be equal, if not superior, to the ones mentioned in the text, has come onto the market. It is called Biolac for Marsupials and is produced in three forms - M100, M200 and M300. M100 is for furless and just furred marsupials, M200 is designed for late lactation marsupials with dark, solid pellet droppings and on solid food. M300 is a more sweet tasting milk, also for late lactation marsupials. The producers of Biolac, Universal Milk Products (P.O. Box 93, Bonnyrigg, NSW 2177. Telephone (02) 823 9874) say that only three species of marsupials have had enough scientific analysis completed on their milk composition that would permit a truly scientifically based substitute milk to be produced. They are the Tammar wallaby, the brushtail possum and the red neck wallaby. "At this stage," the producers add, "Biolac remains a universal milk product for all marsupials."

CHAPTER 2
What To Do

Fostering wildlife is a far more complicated business than raising puppies or kittens. It requires dedication, knowledge, time and emotional resilience. And it's not a job for young children. Marsupials most likely to be found orphaned or injured in Tasmania or, for that matter, the mainland, (apart from the pademelon) are the Bennetts wallaby (red-necked wallaby), rufous wallaby (pademelon), wombat, brushtail possum and echidna. Sometimes, however, you may be confronted with a less common animal such as a ringtail possum, eastern pygmy possum, potoroo, bettong, eastern-barred bandicoot or eastern grey kangaroo (forester).

Before proceeding on what to do if you come across one of these animals it's necessary to outline the regulations as laid down by the Department of Parks, Wildlife and Heritage. The department says that virtually all native wildlife in Tasmania is protected, but qualifies that by stating there are two forms of protection wholly protected and partially protected. (The latter include the two species of wallaby, four species of ducks and deer). Now, to possess wildlife, you need a permit, the only exception being both wallaby species, brushtail possums and wombats that have been taken legally that is, rescued or orphaned. For all other species you need to get in touch with the department for a 'rescue permit'. The department will specify certain conditions, for example, date of release. Some species require special expertise or handling, in which case the department will take over the animal and supervise its rehabilitation. These, typically, are all raptors (owls and birds of prey), the Tasmanian devil and both species of quolls. The department can he contacted at 154 Macquarie St., Hobart, or P.O. Box 44a Hobart 7001 (Ph (002) 33 6556) or at Prospect House, Bass Highway, Prospect or P.O. Box 180 Kings Meadows 7249 (Ph. (003) 36 5312).

What, then, do you do if a young native animal unexpectedly lands in your lap? Its mother may have been shot, or you may see a dead animal on the side of the road with a baby huddled against it. A live joey may even still be in the pouch. (My wife stops to check the pouch of any dead marsupial she sees, regardless of whether or not a joey is obvious). If the joey is naked, its chances of survival are slim. If it has hair, and its eyes are open you may well save it - if you act quickly and follow a few simple procedures.

The first priority is to get the orphan warm and the quickest way is to put it close to your own body. More than likely it will be suffering from shock, so try to keep it quiet, and undisturbed and away from children, no matter how anxious they are to cuddle it.

Also curb any tendency to show the baby off. A herbal product called Bach's Rescue Remedy, available from vets, and most health food stores, helps the youngster overcome initial shock. So, too, does a small amount, of sodium ascorbate (vitamin C) - not to be confused with ascorbic acid - placed on the tip of the animal's tongue. If the animal is completely unfurred it will need, in addition to being covered with a top grade baby oil (Johnston's is the best), frequent small feeds every two hours, day and night. And, I emphasise again, it is absolutely essential that the orphan be kept at a warm, even temperature at all times and that a constant check be kept on the source of the warmth.

As a general rule, a young marsupial will take between one tenth and one fifth of its body weight per day in milk. This means that a 2kg joey having four feeds a day would take 50-100 ml at each feed. (If you can weigh them daily for the first week, then at least weekly - and keep records). Calculate the maximum amount the animal should have, based on its weight, and always try to feed at the same time every day. A successful regime is 6am, 10am, 2pm, 6pm, 10pm and 2am. In joeys that are growing well, the 2am feed can be omitted. Keep the volume of feed constant; variation in amount by as little as 5ml per day may be sufficient to cause diarrhoea. Varying the temperature of the milk can also be a problem. A temperature of about 30 deg C is best. Joeys often refuse to drink milk that is either too hot or too cold.

Initial four-hourly feeds should be diluted to half strength using either boiled water or, for preference, rice water (water in which brown rice has been boiled). If after the first 48 hours, there is no sign of diarrhoea, gradually increase the strength of the formula. As a vitamin supplement, add a drop or two of Apex Vitamin drops to its milk formula. It also helps to overcome stress. If unavailable, use Pentavite which is stocked by chemists.

You may hesitate to carry out what is described next, but it's vital if the orphan is to survive. Marsupial mothers lick the anal area of their young to stimulate the passing of urine or faeces. Now, I'm not suggesting you try this, but the effect can be duplicated by very gently rubbing the area with moist cotton wool or tissue for a few seconds before and after every meal. This will also help relax the animal and keep its bedding clean. Until it has relatively thick fur, its faeces will be, generally, of a thick, pasty consistency. As it develops, the faeces form pellets, and once that occurs you know you're on the path to success. Joy invariably breathes a sigh of relief at the first sight of pellets. They're easier to clean up, too. Some young joeys tend to spill a lot of milk when feeding, so wipe the excess off with a tissue and never put it to bed while it's wet. Check, too, for ticks and fleas when the joey arrives. Ticks, more often than not, are found in its ears.

Should diarrhoea manifest itself, as it often does, feed the animal a mixture of Lectade (a liquid concentrate for oral rehydration therapy) and Gastro Intestinal Absorbent (GIA) both of which are available from vets, Make up a dose by adding one to two teaspoons of GIA to 120ml of Lectade. At the same time dilute its milk formula. As soon as the diarrhoea stops, hopefully within 24 hours, slowly increase the strength of its formula.

If the animal suddenly becomes weak and unable to stand, ask a vet to give it an injection of Dexadresson as well as a dose of VAM (manufactured by RWR) or Aminoplex according to its size. The response is usually quite spectacular. At the same time, keep up the oral Lectade and GIA.

The next stage is when the animal is covered in short, soft baby fur. They should now be starting to eat solid food such as dryish grass, peeled apple slices, orange dandelion leaves and flowers, unsprayed rose and apple leaves, gum tips and, in the case of brushtail possums, dock and Kunzea tips which, we've found, they relish. (Kunzea Baxteri is a native evergreen available from most nurseries). About this time the joey should be introduced to a small bowl containing its milk formula and taught to lap. Once they're lapping, it will save much time and effort, and impose less stress on the orphan.

Don't give them bread or processed food, don't overfeed and don't force feed. This may lead to aspiration of milk and development of pneumonia. As the joey gets older and increases its weight, its dietary requirements will change. Night feeding can be stopped altogether, but make sure, nevertheless, they're fed the last thing at night and first thing in the morning. More solids should be given and more exercise allowed.

If you have reared the orphan to this stage you will begin to appreciate the fact that you have quite a little character on your hands. However, if you feel it's going to be too much of a chore, contact Parks and Wildlife, or the Tasmanian Fauna Society (the secretary is Mrs. E. Jensen, Ph. (003) 44 5358 A.H.) or one of the wildlife parks, and a foster home may be found for it. Some people just do not have the necessary attributes for the task.

Helpful hint: Never feed your wildlife dry dog food. It contains preservatives and could cause diabetes and/or intestinal disorders.

CHAPTER 3
Housing

All new animals should be kept isolated from established animals for a period of time (you must use your discretion on this) to enable them to adjust to captivity and a new diet. In Tasmania it is not against the law to keep partly protected fauna such as Bennetts wallabies, pademelons and brushtail possums. Nor is a permit required, but it is advisable to contact the Department of Parks, Wildlife and Heritage if you plan to keep more than a few.

When constructing an enclosure for wallabies and/or pademelons, bear in mind that they can easily injure themselves. These species take fright at the drop of a hat and will run full pelt into a fence. Fences, therefore, should be easily seen and not likely to result in injuries. Hessian, or shadecloth, attached to the fences can minimise injuries.

While animals acclimatise to captivity, it's important that they are not exposed to excess human activity and that they are screened from prying eyes. Nervous animals should be provided with shelter to hide behind, for example, eucalyptus branches leant against the fence or some bales of grass hay.

For Joy's ever-growing population of pademelons and wallabies, we set aside an area of a little more than half a hectare and enclosed it with a two-metre high wire fence. There were a number of trees growing in the enclosure which we protected by wire guards to prevent the wallabies from ring barking them. There was also a shed, previously used for rearing calves, in which fresh grass hay was spread every few days and which acted as bedding. This area housed eight wallabies of varying ages and 12 pademelons. Apart from a few spats now and then, all got along extremely well together - much to our surprise. A large backyard is sufficient for two kangaroos. To reduce contamination enclosures must be raked regularly, and droppings removed. Remember that some macropodids can become quite aggressive, or even dangerous when they reach adulthood. Pippy, our first wallaby, could have ripped my leg open with a well-aimed kick had I not been wearing heavy jeans. In doing so, he was only protecting his females.

In an urban backyard, a wombat, unless housed in a special enclosure that allows it to dig but prevents its escape, may destroy flower and vegetable gardens, wrecking unbelievable havoc. It is preferable that the enclosure has solid walls, or wire that is unlikely to result in foot injuries - for example, weld mesh.

Provision of a large box lined with hay for a shelter will give some of the security of a burrow.

Bandicoots, being ground dwellers, are best held in completely enclosed cages with a nest-box for security.

Possums are a little more difficult to house. We erected a very large dome-shaped wire enclosure, with a small gate on one side, around the base of a tree and filled it with logs and branches. Possums also need a 'nest box' for security; Joy used a wooden letter box which seemed to suit them admirably. A hollow log is also ideal.

Helpful hint: Don't leave foreign bodies, especially string, lying about. They are common causes of intestinal problems in pet wallabies and kangaroos.

CHAPTER 4
The Red-necked (or Bennett's) Wallaby

The red-necked wallaby is just one species in a diverse - and incredible group of mammals. To confuse things, they are widely referred to as kangaroos, as well as being known by their old name of Bennett's wallaby. Sitting on their long feet, peering curiously at something with which they are not familiar, ears twitching, they are, perhaps, the most photogenic of the. marsupials. Being mainly nocturnal and only partly protected in Tasmania, they are hunted extensively, providing 'sport' for spot lighters. Their colour is normally a light, smoky grey which enables them to merge effectively with their surroundings. They feed on grasses, and like to chew the bark of trees, especially eucalypts. That's why, if they are left in an enclosed area around the house it's advisable to erect wire or wooden guards around any trees that are growing there. Vets believe there is a seasonal correlation (mainly autumn) between bark-stripping and low magnesium levels in certain species of wallabies. In Victoria, provision of commercial magnesium lick blocks led to a significant reduction in bark-stripping. But this behaviour may also be related to hunger for roughage. Pip was our first Macropus Rufogriseus - that's the scientific name for the Red-Necked wallaby. His mum had been knocked down by a car near our farm and he turned up, all legs and hardly any fur, in the bottom of a smelly hessian bag. Joy was delighted, fussed over him like an old woman, wrapped him in a woollen jumper and that night took him to bed. He had the honour of being the first of many to share our sleeping quarters. At this stage, Joy had no idea of what to feed him and Pip survived more by good luck than good management. The first few days were touch and go; Pip developed the scours, was obviously suffering from stress and was reluctant to drink from the special marsupial teat, hastily acquired from the local vet. But he eventually gained a liking for the formula, the scours ceased, and he began to put on weight. From then on, he went everywhere with us - to the shops, on visits, even on weekend trips. He loved the car, travelling in a pouch slung between the front seats. He showed great interest in the scenery, only ducking his head into his pouch when a truck roared past. On one occasion when he was about nine months old we stayed overnight in a small guest house. In the moming the elderly housekeeper decided to make up our room while we were having breakfast. Suddenly we heard a scream. Pip, apparently, had leapt at her from behind the door. We left our breakfast, bundled Pip back into his pouch, paid the bill and hurriedly drove off, to the quizzical looks of the proprietor.

By the time he was three, Pip was the father of several Joeys. But by this time he had become overly protective of his 'wives,' so much so that he became aggressive. (Bennetts can pack quite a kick with their hind legs). It was, we felt, time to release him so, somewhat sadly, we had him tranquillised by a vet and took him in the car to a national park. Our last sight of him was shaking his head, whether from the after effects of the drug or disbelief that he was at last free we don't know, and hopping unsteadily into the scrub. Male handreared Red-Necks, in actual fact, should be castrated if it's proposed to keep them as pets. Pip's behaviour was normal. Fostered animals lose their fear of humans and, when mature, accept them as one of the kangaroo tribe. They will attempt to mate with a woman, as Joy experienced, and fight with a man, as I experienced.

Wallabies, like kangaroos, need a certain level of fibre in their diet for the stomach to function normally. Giving them food such as native grasses and/or hay, fibrous tree bark, gum and wattle branches, seems to reduce the incidence of a fatal disease called lumpy jaw. This is caused by the invasion of bacteria through damaged tissue in the mouth. As a result, the mandibles swell, thus the name. (See also Diseases, Ch.13).

Macropodids also need access to dirt (on the ends of grass roots) to introduce micro-organisms into the gut to form the basis of stomach flora which break up milk particles. Without this flora, persistent scouring (a sign of maldigestion, malabsorption or enteric infection) is likely to occur. If this happens, simplify the diet and dilute the milk formula. Once the animal is back to normal, gradually increase the strength of the formula. Vets discourage the use of antibiotics unless there is good evidence of severe infection.

FEEDING: There are a number of stages in rearing a wallaby: furless, fine fur, thick fur, becoming sure of itself and weaning. The following formula, used very successfully over six years, can be given during all five stages of development: Make up a liquid comprising three scoops of Di-Vetelac to 120 ml of water. Add to this a quarter egg yolk, two drops of vitamin mixture (either Apex Vitamin drops or Pentavite), quarter teaspoon dicalcium phosphate (DCP); quarter teaspoon vitamin E powder, two teaspoons of very fine Farex and half teaspoon Gastric Intestinal Absorbent (GIA). The GIA can, however, be given separately.

All these ingredients are available from vets. The GIA powder is added to prevent diarrhoea, while the vitamin E powder assists in the proper functioning of the gut.

Initially, feed the orphan 6 to 8 ounces per day. By the time it has fine fur, increase the amount to eight to 12oz per day, at the same time allowing it access to grass and dirt. From the time it has thick fur until it shows signs of independence, it will require four 4oz bottles a day. By now it should be on a diet of grass, apple (sliced into small pieces), carrot, lucerne, celery and carrot tops and, if you want to give it a treat, a strawberry or two if they're in season. Special wallaby pellets can be bought from produce stores to supplement the diet. We have found these to be excellent, despite the fact that some vets are hesitant in recommending them. (This is because of claims that the pellets can cause 'black scours,' but we have never experienced this). When it's about eight months old, the joey can be weaned by slowly reducing the amount of milk and replacing it with water. A small amount of Wild Forage, another commercial preparation available from vets, can be added to the water. The joey should, by now, be lapping.

If you follow the above procedures, the joey has an excellent chance of survival, or so we've found. However, Barry Munday* in his report on marsupial diseases, says that one must never dissuade someone who is rearing a joey from continuing with an apparently inappropriate, but successful regime. "These people," he adds, "presumably manage to maintain a normal intestinal physiology in their charges and thus counter the apparent inappropriateness of their feeding regimes. Whether or not cataracts occur in animals raised under these latter circumstances is uncertain, but the high mortalities which can occur when alternative regimes are instituted makes such considerations of minor importance."

---

*	*Munday BL, BVSc MACVSc, Chief Veterinary Pathologist, Mt. Pleasant Laboratories, Tasmania.*

Rufous Wallaby

CHAPTER 5
Paddymelons - 'A Tough Hop'

Pademelons, rufous wallabies, scrub wallabies - they're all the same - always have had a tough hop. When the Thylacine (Tasmanian Tiger) roamed the State, they were one of its main sources of food. Since its demise, they have been stalked by two-legged hunters and poisoned by farmers. "We call 'em stinkies because they smell so much when they're skinned," a farmer informed me, adding that they made "great patties for eating." Back in 1963, Michael Sharland was equally derogative in his book Tasmanian Wild Life.* "indeed," he wrote, "in neither its broad outline nor in its countenance is there anything distinctive or very pleasing. Its fur never seems to be well groomed or nicely combed, or sleek, or of a pleasing colour. The creature is more like a large, untidy rat, the least presentable of the animals that comprise the kindred of the kangaroo. And among its own kind and in its attitude to related species it displays an aggressive and spiteful temper, fighting and tearing out fur. The only good thing that may be said about it is that its flesh is the most delectable of that of any form of kangaroo."

Pademelons are still widespread in Tasmania, although on the mainland, where they were also once abundant, only a few colonies remain. One of our first pademelons was Trixi. She was about three months old when she arrived at the farm and her mother had died of 1080 poisoning, along with 22 others. They had had the temerity to feed on a crop of oats, which had not been fenced and the owner had laid baits. For quite a while I gave up oats for breakfast. The tiny ball of dark brown fur with its furtive eyes would have died too, but she was found by a passer-by in the nick of time, huddled against her dead mother. She was cold, hungry and very frightened. A baby pademelon (their scientific name is Thylogale billardierii) is perhaps the cutest of all young fauna. And after seven years of rearing them, including a blind one, they have become Joy's favourite among the marsupials.

Trixi survived and produced a number of babies. Years later, we had to find a new home for her. From time to time I'd drop in

---

* *Sharland, Michael (F.R.Z.S.): Tasmanian Wild Life - A Popular Account of the Furred Land Mammals, Snakes and Introduced mammals of Tasmania (Melb. Uni. Press 1963) P22.*

to see her, but on one occasion more than nine months had gone by. On entering her enclosure, she recognised me immediately, hopped up into my lap as she used to do when she was small and, uttering a funny 'tch, tch, tch,' sound which, we've learnt, expresses fondness or satisfaction, settled down for a cuddle. Pademelons, like wallabies, express their feelings in a number of different ways. One that is frightened or stressed, for example, will lick its forepaws quite frantically. Tightly clenched paws also indicates nervousness while a cross between a grunt and a snort (the best way to describe the sound) means it's annoyed.

By the time we sold the farm we had 16 pademelons. To our surprise, they got on extremely well with the red-necked wallabies. Occasionally there would be a scuffle among the males, which are larger then the females and have stronger forearms, which assists them in subduing the females when copulating. The young remain in the pouch for eight months.

One of the most important overt diseases of recently captured pademelons and wallabies is called capture myopathy (from myo, the science of muscles). Stress is believed to be the major cause, resulting from more than usual exertion, both during and after capture. Animals acutely affected may die with no signs of illness, other than rapidly progressive depression. Some survive initially, but die within two to three weeks due to a variety of complications. In less acute cases, stiffness, pain and spasm of muscles are exhibited and respiration is laboured. Affected animals should be placed in a soft bag, kept quiet and warm, and given a few drops of 'Bach's Rescue Remedy'.

FEEDING: The formula for pademelons, from the time they have fine fur until they're about 10 to 12 months old is the same as that for Bennett's wallabies. At first feed the joey every four hours. Then, once they appear to be progressing and their tail feels thick and not bony, feed them during the day only, but still at four hourly intervals. In addition to their liquid diet, young joeys can be fed sliced apple and carrot as well as grass, with roots and a small amount of soil attached. Recently wild injured pademelons can be fed carrot tops, blossom and fruit of native shrubs, roses or young leaves from fruit trees, and, of course grass. Wallaby pellets are a good standby make sure they have access to water. This diet will help the animal through the crucial acclimatisation stage.

Helpful hint: Don't leave a pademelon in an area that has no undergrowth or shade. They don't like direct sunlight, and it could damage their eyes.

Bandicoot

CHAPTER 6
The Maligned Little Bandicoot

Another marsupial that has almost disappeared on the mainland, but is still fairly prevalent in Tasmania, is the Eastern Barred bandicoot, also called Long-nosed. It is one of two kinds in the State, the other being the more burly Short-nosed bandicoot. We had a family of EBB's on our farm and at certain times of the year they would come up from the paddock at night and dig around the homestead for corbie grubs. They made a bit of a mess with the lawn, but they ate all the corbies which, if allowed to proliferate, would have eventually destroyed it. Unfortunately, not many people have a good word for the little bandicoot, as they're considered destructive. Ellis Troughton, however, in his book 'Furred Animals of Australia' is one to stick up for them. "They are really useful and much maligned little animals," he writes, "their destruction of mice and insect pests more than compensating for the little damage they may do. Because of their share in preserving nature's balance, as well as their scientific interest, it is unfortunate that members of this family appear to be rapidly disappearing over the whole of the continent." Road traumas are now the most frequent causes of injury and death in these animals.

Now fully protected, the Eastern Barred bandicoot, as the name implies, is distinguished by bands of very pale grey, verging on white, across its lower back and rump. Nocturnal by habit, it raises its young, which can number one to four, in a rear-opening pouch.

In all, there are 16 species of bandicoot (or Peramelids) throughout Australia. Gestation is only about two weeks, pouch life 47 to 55 days and lactation about three months. The young leave the maternal nest about 10 days after quitting the pouch and may first reproduce at four to six months. In their natural habitat - mainly scrub or areas of low ground cover - they feed on insects (especially beetle and cock chafer larvae), birds eggs, seeds, roots and berries.

Orphaned bandicoot are difficult to hand rear and their mortality rate in captivity is high. The young are very delicate and it's extremely difficult to hold them firmly without causing an injury. One I was feeding accidentally fell off my knee on to the ground. It died within minutes. Very gentle handling, then is needed with the young bandicoot. Problems also arise when attempting to keep them in captivity, owing to their solitary nature and the aggressiveness of the males.

FEEDING: Babies and young bandicoots can be fed a mixture comprising eight parts of Wombaroo Insectivore Rearing Mixture to one part of Elliott's Wild Bird Nectar.

If these preparations are difficult to obtain, use either evaporated milk diluted with warm boiled water (one to one) or Di-Vetelac (one scoop of powder to 100ml of warm boiled water). This can be increased in strength as the animal gets older. It's essential to stimulate the urinary vent before and after feeding, as well as between feeds. If the bladder is not voided often, it fills and tilts and can twist, causing death. At one month, the animal's eyes will be open and it will be covered in fur. As it progresses towards the weaning stage, it can be fed small pieces of fruit, small worms and berries. Live insects (crickets) and meal worms; are a preferred food. A diet consisting of only soft foods (milk, bread etc) can lead to periodontal disease, which will prove fatal. It is possible to teach them to lap from a small flat dish. Having said all this, you'll also need luck to rear your little bandicoot.

CHAPTER 7
Brushtail Possums

In 1987, an estimated 252,000 brushtail possums were shot in Tasmania.* Although the figure was slightly less the following year (178,000), the Tasmanian Government "saw no need for an inquiry to be held into brushtail management." In a report, written in 1989, and tabled in Parliament, G.J. Hocking, of the Department of Parks, Wildlife and Heritage, said the species had been harvested for its skins since the earliest times of European settlement and "it now supports a substantial export-based industry." There had also been some interest, he added, in establishing a market for brush possum meat. While on the subject of numbers, the Australian Bureau of Statistics records that during the 10 year period to 1989, a total of 1,260,873 possum skins were exported out of Australia. Luckily for possums, in the past couple of years there has been a 'market downturn' in their skins, "making it hard," one daily Tasmanian newspaper reported, "for those hardy night stalkers with their spotlights and rifles to make a profit."

With so many being killed - there is still a three months 'open' season in Tasmania - more orphaned brush tails arrived on our doorstep, in anything from the toe of a sock to an old rusty kettle, than any other species. What, then, are these marsupials like? Trichosurus vulpecula to give them their scientific name, or "bloody little mongrels" as one person on a talk-back radio programme called them (they had been eating apples in his orchard). Firstly, they're hardy animals that adapt well to many different situations. They suffer from few diseases and respond well to medication. Secondly, when young they love affection and are full of mischief and fun, especially at night, and, thirdly, they make lively companions (but they should be released into a safe area in the wild once they're fully grown). Young brush tails are easy to handle but older ones, especially if they've been injured by a car (again, a common occurrence in Tasmania) must be handled circumspectly.

They are, nevertheless, fairly easily captured by grasping the base of the tail, although care must be taken as the animal can climb

---

*   *Figures provided to the Tasmanian parliament by the Department of Parks, Wildlife and Heritage.*

up its own tail. A serious scratch can result if the handler is not dexterous enough. For closer physical examination, the front claws can be allowed to grasp at an object (a log, or base of a tree, for instance) and the body gently stretched by the tail. While the fore-part of the body is grasped behind the neck while maintaining tension, on the tail, the animal can be safely examined by a second person. It's essential, however, to maintain the tension between the tail and neck. Procedures such as bone pinning, and the wiring of a fractured jaw are relatively simple for vets to carry out and recovery is normally uneventful.

To give an example of how robust these marsupials are, Joy had one she named Sammy. He was in a pitiful condition when he was brought in by the SPCA. There was a weeping ulcer the size of a small saucer on his back, there were open wounds on his forehead and he was badly injured in the back leg. What had happened to him we could only guess; perhaps he had been hit by a car, had dragged himself into the bush and had been attacked by a feral cat. I thought his chances of survival were negligible. Joy, nevertheless, put him into a large wooden box and began to feed him the choicest morsels she could find, including chocolate, which he came to adore, and kept up a regular supply of the Di-Vetelac formula. It took Sammy more than four months to recover: the ulcer healed, leaving only a small scar and although he still dragged his back leg, he could still move surprisingly fast. Then, one night, he managed to enlarge a small hole at the base of his box, pushed away a rock which I'd placed against it and escaped. He must have thought he had had enough of convalescing. One evening, several months later, we investigated a noise in a Banksia tree near the back door of the house; A possum leapt down from the upper branches to within an arm's length; it was Sammy and for the first time he allowed us to stroke his fur without attempting to bite. Perhaps it was his way of saying 'thanks'.

As with other marsupials, baby brush tails respond to close body contact by becoming healthier and more alert. At one stage, when we seemed to be inundated by them, Joy had four in bed with us at night (all in their own small pouches).

As they get older, they enjoy clinging to your arm, just as they cling to their mother's back in their wild state.

In contrast to the American opossum, which is carnivorous, feeding mainly on insects, the Australian possum with its long bushy tail, is mainly vegetarian (although I have known them to kill small birds and chew them). Besides having strong regenerative powers, as Sammy proved, they are extremely resourceful. During a trek through the Walls of Jerusalem in Tasmania's high country, our small party camped one night in a grove of pencil pines. We were busy cooking dinner when one of my companions suddenly exclaimed, "There's a damn possum in my backpack." The possum - it was a fully grown brushtail - had quietly broken into one of the side compartments of the pack, extracted a bar of chocolate (Cadbury's might be interested to know it was one of their's), peeled off the wrapping and was contentedly munching it. Seeing he (or she) had been caught in the act, it leapt up a pencil pine, squatted on a branch just out of reach, and unconcernedly resumed his treat. I stopped my companion throwing a rock at him.

Michael Sharland relates a similar incident.* While camping in a bush hut, be was disturbed one night by a sound beside his bunk. "Reaching for a box of matches on a stool nearby," he relates, "I touched the furry back of a possum, which sprang in alarm to a table scattering tins and cups with a shattering clatter. The lighted match showed that the animal had eaten the last piece of my candle, an article precious by all standards in such a locality. It was not that precautions had not been duly taken against animal marauders; windows and door had been tightly closed and there was no other means of ingress. But this was undoubtedly an experienced prowler and sneak-thief, for it had waited for the fire to go out and had then come down the chimney! And very promptly it went back that way, expedite by the impact of a boot thrown with excellent aim."

Brushtails bear only a single young annually, require about nine months to wean this young and first reproduce from between a year to two years.

Variation in colour is one of their features. The ones Joy has reared, and there have been dozens, have ranged from a very dark brown through to a silver grey. There is also a golden variety, sometimes referred to as the "rock possum" and, very occasionally, an albino is seen. A farmer once told me he had shot an albino- one of the highlights, he proudly claimed, of his life.

Brushtails become unpopular when they 'raid' a vegetable patch or garden. They love new rose shoots, as well as the flowers, and will help themselves to berries and fruit (raspberries are a

---

*   *Sharland, Tasmanian Wild Life, pps 49-50.*

particular favourite). However, we have found that by hanging a small container of dried food (Nutri-Grain breakfast food, for example) from the branch of a tree at night the possums will leave the garden alone and content themselves with devouring the hand-outs. If this doesn't work, quassia spray (made from chips of bark from the quassia tree and available from chemists (hopefully) can be safely used to discourage them. If it's difficult to obtain quassia spray, try sprinkling red pepper around.

FEEDING: You can use the same formula as that for the wallabies for rearing baby brush tails. Another proven formula which is equally beneficial comprises one tin of Bear brand (low fat) milk, one tin (the equivalent amount) boiled water, one egg yolk, two dessert spoons of honey and 12 drops of Pentavite, or Apex vitamin drops, mixed all together. Young babies should be fed every four hours using a kitten or pet teat (regular marsupial teats are too long for brush tails and ringtails). Make sure the volume of food given to the baby is between 10% to 15% of its own weight in each 24 hour period. Thus, if the Joey weighs 200g, it will require 20 to 30ml divided equally between its feed per day. And if the animal is not drinking enough at each feed, it will probably need feeding more often. As they get older they can be fed a varied diet including fruit as wide a variety as possible - unsprayed roses, gum tips, wattle, muesli biscuits, corn flakes, Nutri-Grain and wholemeal bread spread with either honey, peanut butter or Vegemite. For a treat they can be given a square or two of chocolate.

The orphan's progress can be checked with the following age table (supplied courtesy Wombaroo Food Products):

	AGE Days	TAIL mm.	WT. g.
	10	8	4
	15	11	5
	20	15	6
	25	16	7
	30	20	8
	35	23	10
	40	27	11
	45	30	13
	50	35	16
	55	39	19
	60	44	22
	65	49	26
	70	54	30
	75	59	35
	80	65	42
	85	71	49
	90	77	58
	95	84	68
	100	91	79
	105	98	93
	110	105	110
E	115	113	129
	120	121	151
	125	129	178
	130	137	209
	135	146	243
	140	155	282
	145	164	327
F	150	174	390
	155	183	450
	160	193	480
	165	204	525

E = Emerging from pouch. F = Fully Out.

Helpful hint: When orphaned, joeys wash and groom themselves and are liable to swallow their Fur. Brushing them very lightly at feeding time will remove any surplus hair and prevent hair balls forming.

Ringtail Possum

CHAPTER 8
Ringtail Possums - Distinct Characters

Early every morning while Joy was rearing her, Muffin, a baby ringtail possum, would like nothing better than to join us in bed. Scrupulously clean, she would first carefully groom herself, then, ignoring Joy, would make herself comfortable between my legs, stretching herself full-length on her back, open paws pointed upwards, and go to sleep.

Ringtail possums (they are the smallest Australian folivorous marsupial) are delightful animals to rear, although they are prone to nipping if they don't get their own way, or if they become annoyed. Distinguished from the brushtail by their prominent white tipped tail, they became an endangered species in Tasmania earlier this century when countless numbers were shot for their skins. Furriers would use up to 50 skins to make a single coat, while to achieve matching colours, as many as 450 skins would be required depending, of course, on the size of the garment. Now fully protected throughout the country, a permit from the Department of Parks, Wildlife and Heritage is needed to rear them The ringtails Joy has reared all had distinctly different characters. Nippy, the first, was devoted to one person, and one person only - Joy; Honey was clinging and affectionate to everyone, Muffin, with her pranks and quirks, was a sort of possum version of Garfield, while Rambo was forever busy, rearranging his 'nest,' waddling hither and thither, sorting out his food into different piles.

Nippy entered our lives one night when we were taking Callum, our collie, to dog training. A small ball of fur, disorientated by the headlights, darted in front of the car. I hit the brakes and swerved. The wheels just missed him and the little fellow stood shaking in the middle of the road. He was very small and probably had not long left the family nest. If I leave him, I thought, he'll more than likely get run over, so I picked him up and put him under my sweater where he remained contentedly while Callum went through his paces. Nippy stayed with us for about nine months before we felt it was time to release him. Apart from Joy, he would not tolerate anyone else handling him. He would groom her hair and gently nip her neck - a sign of affection. But Heaven help anyone else who tried to nurse him. His favourite dish was small pieces of Ryevita biscuit with honey.

Muffin had a different character. Gregarious by nature, she liked to play games and seemed to possess a sense of fun. We did a fair bit of travelling when Joy was rearing her. Once we stayed at the Sheraton Hotel in Hobart, taking Muffin up to our room overlooking the Derwent in a carry bag. Muffin was duly impressed, especially with the view. After minutely inspecting the room, she took up a position near the broad window, peering through, and refused to budge. In the morning we took her down to breakfast concealed in her bag and slipped her a morsel of bread and honey. For once she kept very quiet.

When angry, ringtails make a throaty 'kee, kee, kee' sound and when they do it's time to watch out as they have exceptionally strong teeth for such a small creature and can move their head very quickly. They enjoy being carried around on one's shoulder.

Known scientifically as Pseudocbeirus peregrinus, the ringtail is smaller and much less robust than the brushtail, which, by the way, will attack and kill them if they wander into their territory. They do not respond well to medication, but there have been few reports of infectious disease in adults. And whereas the brushtail is a wanderer, the ringtail is a nest builder and a 'stay-at-home'. Their nests, usually built on the crown of a tree fern, or in the fork of a tree, are lined with bark, twigs, leaves and grass and they obviously go to great trouble to construct them. The mother normally gives birth to two young which remain in the pouch for about four months. However, she can produce as many as two litters of up to three young annually. Like the brushtail, she'll carry them on her back while foraging.

Cats are one of their greatest enemies, as they are of other small arboreal marsupials. In 1989, one wildlife shelter in Tasmania recorded 312 possums from seven species, including ringtails, which were injured by domestic and feral cats. Of these, 199 died, or required euthanasia.

When rearing a ringtail, construct a nest box and line it with suitable material such as dry leaves and flayed string-bark. This will provide the necessary security for an animal which generally is nocturnal. Perches of box or stringy bark should be included for exercise and developing climbing skills. Ringtails are liable to develop sore claws if they don't get enough exercise.

It will soon become apparent that ringtails are fussy eaters and it can be very difficult to persuade them to eat fruit and vegetables. In the wild they eat shoots and flowers of Eucalyptus, Kunzea shoots, Melaleuca (paper bark), Leptospermum (tea-tree) and acacia (wattles), but these may not always be available. Joy's ringtails, however, developed a liking for walnuts, new rose leaves, native flowers and rock melon. Muffin would tuck into brown bread and Vegemite with gusto. Not one would be tempted with cherries, strawberries or any other berry for that matter. All however, tucked into banana passionfruit flowers.

The following table* is useful in calculating their age and weight:

	AGE Days	TAIL mm.	WT. g.
	10	10	2
	15	11	3
	20	17	5
	25	24	7
	30	32	10
	35	40	13
	40	47	16
	45	55	20
	50	62	24
	55	70	28
	60	77	32
	65	85	37
	70	92	42
	75	100	47
	80	108	52
	85	115	58
	90	123	64
	95	130	70
E	100	138	76
	105	145	82
	110	153	88
	115	160	94
	120	168	100
	125	175	107
	130	183	114
F	135	191	121
	140	198	131
	145	206	153
	150	213	174
	155	221	196
	160	228	217
	165	236	239

E = Emerging from pouch. F = Fully out.

---

*   *Courtesy of Wombaroo Food Products, Glen Osmond, South Australia.*

FEEDING: The formula for baby ringtails is: One tin of condensed milk mixed with two tins of boiled water, plus 20 drops of Pentavite. This amount should last about a week, but if there is any left over, throw it out and make a fresh amount. Follow the same procedure as described for baby brush tails. As the baby gets older, it can be introduced to the foods mentioned above. Vets also recommend a possum nectar mix which should be administered daily. This comprises one litre of water, one tablespoon of honey, $\frac{1}{3}$ litre high protein cereal, $\frac{1}{2}$ level teaspoon Pentavite, and two drops of Copherol E. A mineral salt lick can also be made available, as well as a dish of water. Ringtails should not be fed white bread and jam as problems arise with their teeth and gums.

CHAPTER 9
A Loner - The Echidna

One of the most inoffensive and unique animals in the wild is the echidna, or Tachyglossus aculeatus tachyglossus meaning a sticky or active kind of tongue. Fully protected, these monotremes are solitary creatures and, so far, are not endangered. There were a few on our farm and once in a while one would casually waddle past us while we were sitting on the back verandah in the late afternoon. In the wild they feed by means of a long, sticky tongue on ants and termites, dining on them in the early morning or late afternoon. The females lay a soft-shelled spherical egg and the young hatch after 10 days and are weaned about 10 weeks later. Reports of diseases in echidnas are very rare, but they will die if their snout is severely injured. It is rare for them to breed in captivity.

If you find an echidna on its back, turn it over. It may not be dead, or even badly injured, but just unable to right itself. Because of their spines, you'll need gloves, or some protective hand covering, to handle them. They can be picked up by sliding a gloved hand under the body and grasping the hind leg, but be quick as they can rapidly dig into the ground, becoming almost impossible to remove. R.J. Whittington*•saysthe"surefiremethod of finding a back leg is to touch the animal on its forehead and a hind leg will momentarily shoot backwards." "This reflex will work repeatedly if you miss the first time," he adds. Injuries, sometimes fatal, can occur attempting to dig them out of the ground with a shovel. Another tip: they have an annoying habit of squirting urine and faeces, so point them away from you if you're handling them. You may notice that they blow bubbles through their nose; this is believed to be a mechanism for clearing earth from their nasal passages. If you find an echidna injured, and it needs to be kept for some time, it's best housed in a container without sharp corners. They can hurt their nose on sharp wire and can be difficult to remove when jammed into corners. A large round bucket, or plastic garbage bin, may be suitable but don't carry or leave them in a hessian bag or sack.

---

*   *Richard Whittington B V Sc MACVSc, Veterinary Research Officer, NSW Department of Agriculture, 'Monotremes in Health and Disease'.*

In captivity, echidnas normally accept a range of foods so long as it's been finely ground, or made into a thick paste. Included in their diet should be evaporated milk, minced meat, yoghurt, hard boiled eggs and vitamin and mineral additives. If, by any chance, they turn their nose up at this, you'll have to go searching for ants and termites. I am indebted to Pam Whitney* for the following 'echidna mix': 0.25kg minced beef, three eggs, one litre milk. Heat until almost boiling, then cool. Add ½ teaspoon (2.5g) calcium carbonate, ½ teaspoon Pentavite, 1½ teaspoons (7ml) Nutriderm (1.25ml/animal) 1 tablespoon yoghurt and one or two cups of high protein cereal.

Baby echidnas need to be force-fed on Di-Vetelac for a few days, using an infant's teat. The job requires two people. After a few days, the animal should drink the formula from a bowl. At this stage, offer a custard made by adding 15g of Di-Vetelac powder and one egg to 100ml of water (heated in a double boiler until it thickens). Wombaroo Food Products (P.O. Box 151, Glen Osmond 5064, Ph. 08 272 6077) produce a commercial echidna milk which is available at short notice. Two milks are available, one for animals up to 30 days old and another for those over 30 days.

---

*   *Pam Whitney, B.V.Sc MACVSc, Seminar for Veterinarians (Recent Advance Series) held at Werribee, Victoria.*

Little Pygmy Possum

CHAPTER 10
Pigmy Possums - Good Breeders

There are two other nocturnal species of possums, both threatened by clear felling of forests, that you may encounter - the eastern pygmy possum, whose features include a tail which is naked for most of its length and its pinkish feet, and the little pygmy possum, distinguished by its prominent ears and eyes.

The smallest of these - in fact it's the tiniest of all possums is the Little Pygmy possum (Cercartetus lepidus) which, when fully grown, weighs between six and nine grams. Both species have light, fawny coloured fur and both build nests lined with shredded bark in small holes in trees. They are good breeders. Litters of four or five are not unusual. The young are weaned at about 60 days and they attain sexual maturity at five or six months. They have been known to live for five years.

According to Michael Sharland* the first species made known to science were taken on Maria Island, off Tasmania's south east coast, in 1818 by the French naturalist, Demarest.

In the wild, they feed principally on pollen, nectar, seeds and insects. The Eastern Pygmy possum (Cercartetus nanus) makes a delightful pet, but they are difficult to keep alive. Being wholly protected, a permit is required from the Department of Parks, Wildlife and Heritage to rear them.

If, however, you come across one, place it in a cage and provide it with fresh gum branches twice a week. A warm nest box, lined with hay, fine grass or wool, is essential. Don't use artificial fibres as they can tangle around their toes and cut off circulation; if this occurs, the toes, or even a leg, may have to be amputated. When asleep during the day their metabolism and breathing is greatly reduced and you may think they are sick or perhaps dead. They curl up into a ball so tight that they can even be rolled along the ground! Meal worms, which they like, can be raised in the bottom of their cage and the possum can help itself. They will also feed on huntsmen spiders and large moths.

Sliced fresh fruit (apple, pear, banana) can round off their diet. And, of course, they require fresh water daily.

---

*   *Sharland, Tasmanian Wild Life, P53.*

A successful formula for young pygmy possums is five parts Ellliot's Wild Bird Nectar to one part Wombaroo insectivore rearing mixture. Mix into a paste and leave it in their cage. Pygmy possums tend to eat a large amount and you have to be careful that they are not overfed. If they appear to be getting too fat, cut back on the feed. They eat more in spring, summer and autumn while in winter they tend to hibernate.

Sugar Gliders

Like the Pygmy possum, the sugar glider (Petaurus breviceps) is one of about 41 species of primarily arboreal marsupials, the largest of which is the brushtail possum. The chances of coming across one of these tiny creatures are slim, but if you do, exercise a little caution; they have razor sharp teeth and can inflict a very painful bite. Never play with them.

Sugar gliders are omnivorous, that is, they feed on many kinds of food. In the wild, the diet consists of fruit, seeds, sap, foliage, gum, nectar, pollen and insects. Being colony breeders, they can be kept in small groups, provided that attention is paid to the make-up of the group and surplus males are removed.

Dr. Hume, who has worked on the nutritional physiology and ecology of marsupials for 20 odd years, and who is the author of the book, 'Digestive Physiology and Nutrition of Marsupials', wrote an interesting piece on their feeding habits of sugar gliders for the John Keep Refresher Course for Veterinarians (1988).

The arboreal omnivores, he said, feed on animal material in the form of insects which are high in protein content, and some form of plant exudate high in sugar. "The plant exudates," he continued, "include eucalyptus sap, acacia gum, nectar, manna and honeydew. Manna is a deposit of white encrusting sugars left where sap has flowed from a wound produced by sap sucking insects in a tree trunk or branch. Honeydew is the excess sugar that is secreted by sap-sucking insects. In captivity, the sugary exudates can be replaced by honey and fruit. Insects such as crickets and mealworms, rather than a commercially prepared cat or dog food, are usually provided."

In Tasmania you will need a permit from the Department to rear sugar gliders. They are extremely hard to bottle feed, so try and get them to lap, using the same formula as that suggested for pygmy possums. Grated apple, carrot, muesli and pure baby foods (apple, pear etc) can be given to the growing glider. When they are about half grown, cut out the milk formula and give them honey water. About the same time, begin feeding them on buds, tree saps, native flowers, invertebrates, nectar, green shoots and sweet gum from native plants. A friend who has had considerable experience with these marsupials tells me that they will lose their fur if they play in their milk. If they do get covered in milk, she bathes them in warm water.

As far as housing in concerned, the ideal place, if they are small, is a fish tank, but make sure the lid is on (leaving, of course, space for air) as they will certainly escape. Later they will have to be moved to a larger area (cage, or whatever) with a nest box or a log with a hole in it, as well as branches and plenty of foliage. "Make it really bushy," my friend says. If it's possible, the cage should be big enough for them to glide in.

Wombat

CHAPTER 11
The Common Wombat - An Aussie Battler

In May, 1990 The Australian newspaper published a letter from a couple living in Victoria, who signed themselves "Two of the Privileged Few." The letter is well worth repeating as it is a heart-felt SOS for that unusual and uniquely Australian marsupial, the wombat:

"After many years of caring for wildlife, we believe that our timid and lovable wombat symbolises the Aussie battler. Because this little Australian is slow, territorial, trusting and not aggressive, and asks for nothing more than a hole, some scrub and protection from low-mentality humans and dogs, human Australians treat our wombat with contempt and cruelty. We have had them delivered to us mutilated and sometimes beyond help because they have been butchered alive by a machete-wielding protea farmer, impaled with a steel fencing post by a carrot farmer, snared and left to suffocate by sheep farmers, poisoned by foresters and left to suffer for days by unthinking motorists. We know of many, however, who have wombats as pets. As they make the most lovable, companionable, loyal and low cost and low care pets, it is time our laws were changed so that we can all know the pleasure of enjoying our unique and wonderful but fast disappearing wildlife heritage. As the law stands, all can maim and destroy our wildlife with impunity, but only a privileged few can experience their affection."

As the writers* state, wombats will readily reciprocate kindness with love, sometimes bordering on possessiveness. I have known a wombat to make heart-rendering cries when taken away from its foster parent. One of the wombats Joy reared would doggedly amble after me over the farm for hours on end, studiously ignoring any distractions like curious cows.

The common wombat (vombatus ursinus) has been kept in captivity for more than 100 years. And not only in Australia. Dante Gabriel Rossetti, a London artist who lived during the last century, was particularly fond of wombats and referred to them as "a Joy, a Delight, a Madness" and found them "engagingly lumpish." One wombat he kept "slept in the bowl of a large hanging lamp

---

*   *Maudie and Ralph Shackleford of Emerald, Victoria.*

and once disgraced itself by eating the hat of Mrs. Tebbs, a solicitor's wife, who was sitting for her portrait. It died after only two months and was stuffed, thus continuing posthumously to greet visitors as they entered the hall."

This, burly, rotund animal, which has been likened to a small bear and is called a badger in Tasmania, is mostly nocturnal. It has a rear which is an extremely tough dense mat of skin and fur that acts as a defence mechanism by crushing would-be predators - for example over zealous dogs - against the wall or ceiling of its burrow. These burrows can be up to 20 metres long, with connecting tunnels. Studies have revealed that the majority of young are born in June and July and don't leave the pouch permanently until they are about 9½ months.

Wild wombats are most likely to be injured by cars or by gunshot. A large, aggressive animal - they are armed with strong teeth and claws - may be simply captured by encouraging it to charge while holding a large hessian bag. Remember, however, to quickly step aside, allowing the animal to run headlong into the bag. If relatively quiet, they can be approached from behind, grasped around the body immediately behind the front legs and lifted up against the chest for examination or carrying.

Baby wombats, on the other hand, are easy to handle and to rear if they are more than 500g in body weight. Even if they don't have any hair, they have a good chance of survival - unlike other marsupials. Once the first teeth have erupted, they can be introduced to fresh grass and chopped carrot. The joey should also be kept at an even temperature of about 32 deg. and hairless babies should be regularly massaged with baby oil. Feeding hairless orphans should be at about two hourly intervals, but once fully haired this can be extended to once every four hours.

By the time they are three-quarters grown, provision should be made for them to burrow, for although they make wonderful pets, having a wombat loose in the house at this age is tantamount to having Whelen the Wrecker move in.

Wombats appear to suffer from few diseases, but one condition they are prone to is sarcoptic mange, a skin disease also seen in brushtail possums. In the wild state it may cause death by restricting the animal's vision and movement. In captivity it can be successfully treated with injectable Avomec. However, do not use except under the supervision of a vet, who has accurately weighed the wombat before it is treated. And don't cuddle an affected animal, as transmission to humans is likely to occur.

Being basically herbivorous, the diet of an adult wombat in the wild consists of roots, leaves and grass; in captivity, the fully-grown animal c an be fed fruit, lucerne or meadow hay, horse flakes and wallaby pellets (available from produce stores). As well, give them gum and wattle branches and an occasional supplement of fresh grass. A mineral and vitamin supplement may also be necessary. Put the food in an elevated tray, or trays, (about 20cm above the ground) to prevent the animal defecating and urinating on it.

A note of caution: Don't leave fertilisers, especially urea preparations, around as they tend to attract wombats. Urea poisoning can be fatal for them, and very distressing for the foster parents.

FEEDING: Wombats have decided likes and dislikes when it comes to milk formulas. What one animal will enjoy, another will reject. However, Di-Vetelac (three scoops to 100ml) together with a small amount of Wild Forage and GIA seems to be the most popular. Wombaroo also put out a Wombat mixture. If you're stuck, use evaporated cow's milk, or goat's milk, diluted with warm water. When first using evaporated milk, dilute it with four times the volume of water. A 1kg animal should be given between 15 and 20mls every four hours through a small teat - a baby's teat will suffice. Strict adherence to a feeding schedule and to the volume fed every time is most important. Small joeys should be kept out of their substitute pouch only during feeding time as, like other young marsupials, they cannot regulate their body temperature and hypothermia may result. When the animal weighs about 2kg, the volume of milk given at each feeding can be increased to 30ml every four hours. Once it has become well covered with fur, it can be trained to drink from a dish. If there is any sign of diarrhoea, give the animal Lectade and GIA. (To each 120ml of Lectade add 1-2 teaspoons of GIA). Liquid paraffin added to the diet controls constipation. At 2kg, the youngster can be introduced to solid foods. In addition to chopped carrots, crushed maize, muesli containing dried fruit and sliced apple can be given.

CHAPTER 12
Bettongs and Potoroos

So far I have dealt with the more common species of wild life with which you are likely to come into contact. There are many others, but two that should be mentioned are the small macropods, the bettong and the potoroo (both these names are of Aboriginal origin). Like the larger macropods (kangaroos and wallabies) they are most likely to be injured by a car, or by gunshot. The long-nosed potoroo - there's also a rare long-footed variety - is found throughout Tasmania, Victoria, NSW and King and Flinders Islands. Both it and the bettong are fully protected.

While the potoroo could at first be mistaken for a pademelon with its red-brown fur, the bettong is distinguished by its fawny-grey fur over its body, except its underneath parts which are lighter, its round ears and (usually) its white tipped tail. Another feature of the bettong is its relatively long hind foot, larger than its head. Quarrelsome and pugnacious, it readily fights with its own kind, their bellicosity being equalled only by the ferocity exhibited when two fully-grown brush tails do battle.

Commonly known as the 'pig-rat' or 'wallaby-rat' the potoroo is a nocturnal creature, hiding itself in a nest of grass and leaves usually built in a depression in the ground, during the day. Michael Sharland says it's unfortunate these attractive little marsupials are not endowed with names that might ensure for them the patronage of popular favour, to which they have a genuine claim."More often in the public mind they suffer the odium of being some sort of giant rat, but we need only to get a good look at them to realise they are the most graceful and most delicately formed of our native 'kangaroo' tribe."

FEEDING: Potoroos and bettongs are omnivorous. Recuperating in captivity, they can be given a spadeful of soil, with fallen leaves and roots, insects, fungi, peanut butter, honey on bread, pieces of lettuce, meal worms for older animals, and some finely chopped cooked chicken.

If the animal has no hair, feed every 2 to 4 hours; finely furred, 4 to 6 hours; and well furred, 6 to 8 hours. As soon as teeth emerge, or they show an interest in leaving the pouch to forage, offer them some solid food. Joeys that take only small amounts at each feed need to be fed more often. For joeys, make up the following formula: to each 120mls Di-Vetelac add one teaspoon Wombaroo Insectivore Mixture. Then add ¼ egg yolk, 1ml Wild Forage, ¼ teaspoon Vitamin E powder, ¼ teaspoon DCP (Di-Calcium Phosphate) and 2 teaspoons of fine Farex.

40

CHAPTER 13
Other Common Diseases and Problems.

Cats, or rather cats' droppings, are the bane of marsupials. Kangaroos, wallabies (both Bennetts and Rufous) and wombats are particularly susceptible to a parasitic disease called toxoplasmosis. In most cases it's fatal with no obvious signs that the animal has been suffering. Sometimes, however, the animal may develop difficulty in breathing and in trying to co-ordinate its movements. Loss of balance follows and it may become partially blind. Several of Joy's earlier joeys fell victim to the disease and she despaired, for a time, as to what it was and what caused it.

Both domestic and feral cats are the trouble. Infective sporocysts are passed in their droppings and marsupials are prone to pick them up while grazing in infected areas. Research into a cure has been hampered by lack of funds. Sometimes a vet will suggest using Tribrissen to counteract the disease, but it's rarely successful and it's costly as well. Make sure, then, that cat faeces don't contaminate any feed containers or areas in which the marsupial is allowed to graze. Even bedding used by a cat should not be placed near the animal.

When dealing with diseases, it's most important to have the animal first checked by a vet for worms and coccidiosis, the latter being a disease caused by an internal parasite, and have a faecal test done before treatment. Worms can be cured by using Panacur 2.5 or Fencare 2.5 - In severe cases, repeat the treatment after six weeks. Coccidiosis can be treated with Amprolmin plus (0.6mls per kilo body weight) to be given orally.

Most difficulties associated with hand-rearing native fauna can be traced back to bad management, for in the wild they're pretty tough little creatures. Two very common problems are dehydration and diarrhoea (which has been mentioned in previous chapters). Dehydration can be caused by the animal not taking enough fluid, overheating, or from diarrhoea. If it isn't treated promptly, the animal will die. A simple test is to gently pinch up the skin between the shoulders. If it stays up in a wrinkle and falls back slowly, the animal is dehydrated. Another tell-tale sign is dryness of the mouth. To alleviate this, give the animal Lectade, a liquid concentrate for 'oral rehydration therapy' which is available from vets. Easy-to-follow directions are on the bottle.

Causes of diarrhoea include incorrect formulae, sugar, shock and chilling. In many cases this can be alleviated by diluting the milk formula, or even replacing it with a mixture of 5-10 gm of glucose and 1gm of electrolyte powder (Trolyte) in 100ml of boiled water and administered for up to 48 hours. If severe diarrhoea occurs, a mixture of Slippery Elm bark powder, G.I.A. mix (gastro intestinal absorbent) and Lectade is recommended. In very severe cases of diarrhoea or bacterial infections, the animal should be given Chloramphenicol injections by a vet. Always replace bacteria with Lactobacillis Acidophilus. When the diarrhoea abates, start increasing the strength of the milk formula gradually over the following week. Try the animal with dandelion leaves and roses. Dandelion is a natural astringent and should be used in preference to manufactured products such as kaolin and pectin. Vets discourage the use of antibiotics unless there is good evidence of a bacterial gut infection. For any mild infections, use Sodium Ascorbate powder.

When chilled (a likely state for a newly-orphaned joey to be in) the animal could rapidly suffer hypothermia or hypoglycemia. They'll become listless or depressed or even comatose if they don't receive nutrition at regular intervals. Warmth, and the administration of glucose, plus electrolytes, will usually bring them around. Often, when the animal is chilled, it is also suffering from shock. Ask the vet to give it Dexadresson and 'VAM' (or Aminoplex)

Another disease that occurs in kangaroos and wallabies is thrush which is caused by yeast. Symptoms are sore mouths and refusal to drink. The lips may become swollen with curd-like encrustations developing on the gums and tongue. As soon as these signs appear, give the joey three to four doses each of 100,000 units of Nystatin daily (each ml contains 100,000 units) Nystatin is a yeast inhibiting preparation, also given to humans. Treatment should be continued until the encrustations disappear.

Already mention has been made to lumpy jaw, a disease of captivity or confinement associated with kangaroos and wallabies. It is another killer. The disease affects the oral cavity of the animal and is recognised by swellings of the face and jaw and discharges from the nasal passages and eyes. Victims become depressed and anorectic; there is a rapid weight loss and death follows quickly. according to Dr. J. Burton* treatment is notoriously unsuccessful as recurrence is usual.

---

*   *Dr. John Buron, Chairman, Seminar For Veterinarians (Recent Advance Series) held at Werribee, Victoria.*

Studies have revealed that unnatural dietary regimes produced pathological changes in the periodontal tissues, rendering them highly susceptible to infection. To prevent what must be a very painful disease, make sure any enclosure housing kangaroos or wallabies is kept clean and free from their droppings. Also ensure they are provided with natural feeds, such as grass hay etc. On no account feed them bread. If an animal does become infected, isolate it immediately and contact your vet. It is possible an effective vaccine may be developed in the near future.

Diseases in marsupials would be a sizable book in itself. I've only listed a few that you are most likely to encounter. But while on the subject of what can befall your joey, make sure you know where it is grazing when you allow it outside. Gardens are full of dangers. Defender and Baysol snail pellets are deadly. Rhubarb, rhododendrons, oleanders, azaleas, deadly night shade (ink plant) and yew trees, to name a few, are all poisonous. So, too, are most indoor plants, so keep the animals away from them.

In conclusion, injecting against tetanus, rife in some areas of Australia, should be considered, but consult your vet.

CHAPTER 14
What not to feed

The following items should not be fed to marsupials as they are all toxic*.

Common Name	Scientific Name
Comphrey	Symphytum sp.
Heliotrope	Heliotrope sp.
Cynoglossum	Cynoglossum sp.
Bracken Fern	Pteridium aquilinum
Iron Wood	Erythrophloeum chlorostachys
Gidgee	Acacia georgina
Black Nightshade	Solanum nigrum
Winter Cherry	Solanum pseudo capsicum
Kangaroo Apple	Solanum aviculare
Green Potato	Solanum tuberosum
Rhododendron	Andromedia sp.
Apple seeds	Malus sylvestris
Loquat seeds	Eriobotrya japonica
Quince leaves	Cydonia oblonga
Native Couch	Brachyachne convergens
Sorghum greens	Sorghum sp.
Linseed cake, meal	
Cabbage	Brassica sp.
Sour Sob	Oxalis per-caprae
Sorrel	Rumex sp.
Variegated thistle	Silybum Marianum
Cape Weed	Arctotheca calendula
Hemlock	Conium maculatum
Black walnut	Juglars nigrum
Oak	Querus sp.
Pine tree	Pinus radiata
Oleander	Nerium oleander
Red Maple	Acer rumrum
Almond, Apricot	Prunus sp.
Cherry, plum, peach	Prunus sp.

Rhubard leaves and turnip tops can also be added to this list.

---

*   *List prepared by Pam Whitney B .V. Sc. M.A.C.V.S. Seminar for Veterinarians (Recent Advance Series).*

CHAPTER 15

Releasing

The time inevitably arrives when your native animal should be released back into the wild. Hand-reared animals and ones held for lengthy periods in captivity require step-by-step acclimatisation (natural food, housing, fitness, temperature adjustment) and must develop normal behavioural patterns such as food seeking. This my take several months and should always be kept in mind during rearing and the time held in captivity. It is often a long-term disadvantage to the animal to make pets of wildlife if eventual release is the goal. Sometimes, an animal will be most reluctant to go. It occurred when we released our ringtail, Honey. She kept running back after us. Ellen Jensen's* brushtail, Penny, was another example. "Penny came to me as a tiny, unfurred baby," she relates. "I had a real battle to stabilise her digestive system and clear up patches of raw skin, but at length managed this with minute doses from a huge pink tablet used for bovine gastric upsets. When she was about a year old, she decided to be free and tunnelled out from under her cage. Fair enough, I thought, she'll go into the trees at the back and make her own way in life. But it was not to be. The first few nights she came bounding down from the trees, over the workshop roof, looking for tit-bits and a cuddle. I received a very comprehensive face wash, and my hair was rearranged possum fashion. The first Saturday night I had to go out, and I thought she would be sure to give up and finally go. But no. When I came home at midnight, there she was, waiting on the back door mat. She has now moved on to greater achievements. She has discovered windows through which she can get into bedrooms and, having gained entry, has found there are warm beds into which one can crawl. I know I'm hallucinating; this cannot be happening. Native animals, I'm told on the best authority, do not become tame and bond to their foster parents. And certainly do not return when set free. It's just that hysterical emotional women become attached to them! That being so, I've reared a most unusual possum."

Possibly Honey and Penny were exceptions. Most of the animals we've released have gone off, presumably, quite happily.

---

*   *Mrs. Ellen Jensen, Secretary, Tasmanian Fauna Society, writing in the Society's newsletter No. 12.*

However, there are a few things to keep in mind. On no account, don't just dump the animal, whatever it is, in the nearest bush. Its chances of survival would be slim. Choose an area well away from traffic and, if possible, near the site where the animal was originally found.

In the case of injured animals, make sure they are well and truly fit before releasing them back into their accustomed habitat. Minor injuries will heal better in the bush, and the quicker they are returned the better as the stress of captivity may aggravate their condition. If the animal has been in captivity for some length of time, it will have to re-acclimatise itself, so a-gradual release is advised. If it's possible take their 'home' (box, shelter or whatever) to a suitable site, leaving the animal in it for a few days, at the same time providing adequate food and water. Then open the door, still leaving food in the home until they no longer use it, or are known to be coping well. In the case of possums, I've known them to return to their 'old home' for at least a few days before quitting it for good.

Hand-reared orphans have the best chance for successful release if they belong to a species which are largely instinctive, or have had minimal contact with humans once out of the pouch. It is, therefore - and this is one of the most difficult aspects of rearing - important not to become too attached to them. (Of course, this is much easier written than done). Nocturnal species, like possums, should be released at night fall. And avoid releasing male macropodids during the breeding season - usually summer.

Some hand-reared joeys may be too tame to release. Unaware of the ways of the world and the fact that there is no compassion in nature, they are likely to be shot, suffer exposure, run over or become easy prey for predators. We learnt a sad lesson with Snuggles, one of Joy's brush tails. From an ugly, scrawny, almost hairless baby, suffering from the scours and with a protruding anus, he became one of the most beautiful of bush animals, with soft, silver grey fur and an unusual brownish-white tipped tail, almost like a ringtail. When given the opportunity he would sit contentedly on Joy's shoulder, gently nibbling her ear lobe. We took him with us when we sold the farm and, believing him to be safe at our new home adjoining a national park, released him. One morning, three days later, we found him dead on the road at the back of the house.

Ringtail possums, being more delicate, are more of a problem. The Department of Parks, Wildlife and Heritage has 'half-way houses' to assist them through their pre-release stage, but you may be able to discover where a colony lives and release the animal in that vicinity. A wildlife park, or a vet, may be able to help.

Wombats also need to be released in an area where there are other wombats, but not in the middle of a large group of them. They do fight among themselves and can inflict serious injury on one another.

In conclusion, all hand-reared wildlife need to be weaned off people, as well as the manufactured food they're given. If you have reared an animal and it cannot be released (unless you have a proper 'sanctuary') then you have failed in the most important part of its rescue. However, difficult as it is, the reward in rearing wildlife is the knowledge that a unique animal has been saved from an untimely end and returned to its proper home. To get to know these ancient marsupials from the time the Australian continent was part of Gondwana land, and to be the recipient of their affection, is an unforgettable experience.

CHAPTER 16
A Final Word.

Two years ago Dr. Ebbe Nielsen, a CSIRO scientist and director of the Australian National Insect Collection, was quoted as saying that the world is witnessing the greatest period of extinction in history - a rate that is faster than at any other time in the past. In Australia alone, it's estimated that about a quarter of our mammals perhaps 60 to 70 species - are endangered, and another 100 or so could be at risk.* Since Captain Philip raised the Union Jack at Sydney Cove in 1788, 18 mammals and at least one bird have become extinct - that's an average of nearly one creature lost for every 10 years of European settlement (not counting insects, plants or other flora and fauna). According to the book Australia's Endangered Species** this is higher than in any other continent and country and threatens the very foundations of the evolutionary process. The list is somewhat contentious, put together on the basis that no evidence of credible sightings has emerged in the past 50 years. For the record, it comprises: the pig-footed bandicoot, desert bandicoot, lesser bilby, broad-faced potoroo, crescent nailtail wallaby, eastern hare-wallaby, central hare-wallaby, toolache wallaby, lesser sticknest rat, white-footed tree rat, Alice Springs mouse, Goulds mouse, short-tailed hopping mouse, long tailed hopping mouse, Darling Downs hopping mouse, big-eared hopping mouse, desert rat kangaroo, paradise parrot and the thylacine, or Tasmanian Tiger. Quite a deal of controversy exists over the thylacine as there have been quite a number of alleged sightings over the past decade, including one by a friend of this writer who, at the time, was trekking in a remote area in south-west Tasmania. He is positive it was a thylacine.

Many of these animals were abundant before their habitat was destroyed. Fortunately, the marsupials listed in this book are still common and indeed, some have benefited from land management practices. But others, like the bettong, are losing, through forestry and agricultural practices, including land clearing, large areas of natural habitat. All, however, are threatened not only by feral cats (if not physically, then through the ravages of toxoplasmosis

---

*   *Weekend Australian newspaper, Sept. 1-2, 1990.*
**   *Australia's Endangered Species, edited by Michael Kennedy and published by Simon and Shuster.*

passed in their droppings) but by 1080, a highly toxic poison, laid liberally in Tasmania, which causes an agonising death.

Unless certain attitudes change, including the long held belief in some quarters in the 'right' to kill our 'common' animals will, in time, disappear too.

REFERENCES

AUSTRALIAN WILDLIFE - a series of lectures given at the John Keep Refresher Course for Veterinary Surgeons at the University of Sydney; in particular. Barry Munday who spoke on Marsupial Diseases, Tessa Frazer-Oakley, E.P. Finnie and R. Speare.

SEMINAR FOR VETERINARIANS (Recent Advance Series), held at Werribee, Victoria, under the chairmanship of Dr. John Burton.

MANAGEMENT AND DISEASES OF YOUNG MARSUPIALS, by David Obendorf, Veterinary Pathologist, Mt Pleasant Laboratories, Launceston.

TASMANIAN WILD LIFE, by Michael Sharland.

TASMANIAN MAMMALS, a Field Guide by Dave Watts.

WOMBAT: Maintenance in Captivity, blood values, infections and parasitic diseases, by P.J.A. Presidente.

HAND RAISING OF ORPHANED KANGAROOS by Geoff and Christine Smith, Mt. Pritchard, NSW.

SOUTHERN HAIRY-NOSED WOMBAT, its maintenance, behaviour and reproduction in captivity, by M.D. Gaughwin.

NOTES

NOTES

NOTES